The EV batteries Evolution: Putting The World On High Alert

Michael Kante

All rights reserved. No part of this publication may be reproduced, distributed, or transmitted in any form or by any means, including photocopying, recording, or other electronic or mechanical methods, without the prior written permission of the publisher, except in the case of brief quotations embodied in critical reviews and certain other noncommercial uses permitted by copyright law.

Copyright ©Michael Kante , 2022.

CHAPTER 1; What are EV batteries

CHAPTER 2; How are EV batteries manufactured?

CHAPTER 3; Types of EV batteries and how do they work?

CHAPTER 4; The Future of EV Battery And its worldly impacts.

CHAPTER 5; Conclusion

CHAPTER 1; What are EV batteries

An electric vehicle battery (EVB, sometimes known as a traction battery) is a rechargeable battery used to power the electric motors of a battery electric vehicle (BEV) or hybrid electric vehicle (HEV) (HEV). Typically lithium-ion batteries, are specially developed for high electric charge (or energy) capacity.

Nissan Leaf cutaway showing part of the battery in 2009

Electric vehicle batteries differ from starting, lights, and ignition (SLI) batteries since they are meant to supply power over an extended period and are deep-cycle batteries. Batteries for electric cars are distinguished by their relatively high power-to-weight ratio, specific energy, and energy density; smaller, lighter batteries are preferred since they lower the weight of the vehicle and hence increase its performance. Compared to liquid fuels, most contemporary battery technologies have substantially lower specific energy, and this typically limits the maximum all-electric range of the vehicles.

The most prevalent battery type in current electric cars is lithium-ion and lithium polymer, because of their high energy density relative to their weight. Other types of rechargeable batteries used in electric cars include lead–acid ("flooded", deep-cycle, and valve-controlled lead acid), nickel-cadmium, nickel–metal hydride, and, less often, zinc-air, and sodium nickel chloride ("zebra") batteries. The quantity of electricity (i.e. electric charge) stored in batteries is measured in ampere-hours hours or coulombs, with the overall energy commonly measured in kilowatt-hours (kWh) (kWh).

Since the late 1990s, developments in lithium-ion battery technology have been driven by demands from portable devices, laptop computers, mobile phones, and power tools. The BEV and HEV marketplace has received the advantages of these developments both in performance and energy density. Unlike older battery chemistries, particularly nickel-cadmium, lithium-ion batteries may be drained and recharged every day and at any level of charge.

The battery pack makes up a considerable expense of a BEV or an HEV. As of December 2019, the cost of electric car batteries has reduced 87% from 2010 on a per kilowatt-hour basis. As of 2018, cars with over 250 mi (400 km) of all-electric range, such as the Tesla Model S, have been marketed and are now available in several vehicle sectors.

In terms of operational expenses, the price of electricity to power a BEV is a small fraction of the cost of gasoline for identical internal combustion engines, indicating improved energy efficiency.

Practical electric cars debuted throughout the 1890s. An electric car retained the automotive land speed record until roughly 1900. In the 20th century, the high cost, limited top speed, and short range of battery electric cars, compared to internal combustion engine vehicles, led to a worldwide fall in their use as private motor vehicles. Electric vehicles have continued to be employed for loading and freight equipment and public transport - notably rail trains.

At the beginning of the 21st century, interest in electric and alternative fuel vehicles in private motor vehicles increased due to: growing concern over the problems associated with hydrocarbon-fueled vehicles, including damage to the environment caused by their emissions; the sustainability of the current hydrocarbon-based transportation infrastructure; and improvements in electric vehicle technology.

Since 2010, combined sales of all-electric cars and utility vans achieved 1 million units delivered globally in September 2016,[1] 4.8 million electric cars in use at the end of 2019,[2] and cumulative sales of light-duty plug-in electric cars reached the 10 million unit milestone by the end of 2020.

The global ratio between yearly sales of battery electric cars and plug-in hybrids grew from 56:44 in 2012 to 74:26 in 2019, ad declined to 69:31 in 2020.

As of August 2020, the fully electric Tesla Model 3 is the world's all-time greatest selling plug-in electric passenger car, with roughly 645,000 units.

CHAPTER 2; How are EV batteries manufactured?

Batteries are concealed from view, yet they are one of the biggest and most critical components of any electric automobile.

In this post, we'll look at what electric vehicle (EV) batteries are comprised of, how they store their energy, and how the technology and ownership model for batteries could change. Fully knowing your electric vehicle helps you get the most from it, so read on for our full guide.

What are EV batteries composed of?

The majority of EVs employ lithium-ion batteries, comparable to those in consumer goods such as laptop computers and smartphones. Just like a phone, an electric vehicle battery is charged up using energy, which then is utilized for power, in this instance to operate the automobile.

Whereas the batteries for most electronics have a fixed duration before they are exhausted, EV batteries have a 'range' – i.e., a distance that the vehicle can go before the batteries run out of power. After this, they will require recharging again.

Electric vehicle batteries are not a single unit but comprise hundreds of cells. As a rule, a bigger number of cells often signifies a larger capacity battery, and consequently a longer range of miles that the automobile can drive.

Hybrid automobiles frequently employ nickel-metal hydride batteries rather than lithium-ion. Their lengthy life cycles, safety, and resistance to abuse make them ideal for hybrid vehicle producers. However, they have a high cost and may lose heat or discharge themselves at high temperatures.

In contrast, lithium-ion battery technology has a high energy density and is suited to short charging cycles — perfect for an electric automobile. It also preserves that energy density across hundreds of such charge cycles.

How are EV batteries made?

EV batteries are created from a mix of raw components. 'Base' metals such as aluminum, copper, and iron are important ingredients, but the most expensive materials are 'precious' metals such as cobalt, nickel, and manganese, along with elements such as graphite and lithium.

These materials have to be extracted or mined from the earth in a complicated and expensive way, which is one reason why electric vehicles are more expensive to buy than combustion cars. The most expensive component of an EV is its battery.

The extraction of these elements is regarded by some to be contentious. Many such metals are exclusively found in particular parts of the globe, such as China and South America. Mining them may lead to challenges with international politics and supply chain domination, as well as humanitarian considerations. Lithium also requires a lot of water while being mined, presenting possible complications for agriculture.

Unfortunately, comparable ethical difficulties occur over the globe with the manufacture of fuel and diesel. Crude oil extraction in Eastern Europe may also be susceptible to political, financial, and environmental conflicts, for example, and the frequent swings in gasoline prices might reflect this.

In terms of local emissions, especially if they are recharged from renewable energy sources, electric cars are typically viewed as being considerably better for the environment.

A survey from the EEA indicated a typical electric vehicle in Europe generates fewer greenhouse emissions and air pollutants compared with a petrol or diesel counterpart.

Emissions are normally greater in the manufacture of electric automobiles, however, they are frequently compensated by reduced emissions throughout the use cycle.

Volkswagen is one automobile company that is creating sustainable battery manufacturing utilizing renewable energy sources. This will become more widespread as growing numbers of enterprises invest money and attention into environmental manufacturing practices.

Even if the electricity used to charge an EV has been generated partially by burning fossil fuels, the increased volume and weight of these power plants allow them to be far more efficient than the internal combustion engines of petrol and diesel vehicles which are constrained by both size and weight.

Are EV batteries better for the environment?

How will EVs transform our roadways and affect the environment?

The route to electric — in graphics and statistics

How dependable are EV batteries?

Electric (and hybrid) automobiles have proved to be among the most dependable vehicles on the road. This is backed up by extensive guarantees for EV batteries, which often surpass the whole manufacturer warranty for the vehicle (eight years or 100,000 miles is normal) (eight years or 100,000 miles is typical).

Just like the lithium-ion batteries in consumer electronics, EV batteries do deteriorate over time and during repeated charge cycles, although the drop-off in performance is significantly less severe than in many smaller electronic devices, such as smartphones. That's because the number of charge cycles is significantly more frequent in these gadgets than in an EV; a lot more critical element in battery deterioration than age.

Things you can do to conserve your EV battery.

Keep the charge level between 20 percent and 80 percent to help your battery last longer, and don't constantly use DC quick chargers, since this might have a detrimental impact on battery life. Read more recommendations in our guide to how long EV batteries last.

How does the battery affect EV range and performance?

Rather than the number of horsepowers it possesses, the power output of an EV is preferably measured in kW (kilowatts) (kilowatts). The battery's energy storage capacity is measured in kWh (kilowatt hours) - somewhat like the number of liters a traditional car's gasoline tank can carry.

EV batteries are costly, therefore it's frequently the most pricey vehicle that provides the maximum performance and the longest range. The lowest capacity batteries tend to be in the smallest automobiles, and vice versa.

As EV battery technology advances and grows better all the time, ranges are growing. Mercedes-Benz has just crossed the 1,000km (621 miles) barrier with its VISION EQXX concept vehicle.

It's worth noting that bigger batteries don't always translate to longer charging times. More expensive EVs tend to have fast-charging capabilities that readily compensate for a greater capacity battery.

Electric vehicle batteries are relatively hefty, which might impact how a car handles. However, to mitigate this, they are normally kept under the car's floor which offers a lower center of gravity to help to handle. This has another advantage, too. The 'skateboard' chassis – so-called because when seen from above, the chassis resembles a skateboard with the battery in the center and the wheels at each end — liberates greater internal space.

CHAPTER 3; Types of EV batteries and how do they work.

As the market for EVs increases and customers have more options when it comes to manufacturers and models, it's crucial to understand the various sorts of batteries and what goes into them.

According to the U.S. Department of Energy, there are four basic kinds of energy storage devices utilized in EVs:

Lithium-ion batteries: These are utilized in gadgets like smartphones and laptops. They are commonly chosen in the EV sector due to their high efficiency, strong performance at high temperatures, minimal self-discharge (a chemical process that causes loss of charge), and the fact that most of their components can be recycled.

Nickel-metal hydride batteries: You may find them in many hybrids on the market, however in most plug-in EVs, they have been substituted by lithium-ion batteries. The primary issues with nickel-metal hydride batteries are their high cost, excessive self-discharge, and limited performance at higher temperatures. However, they are typically regarded as safer than li-ion batteries, owing to the absence of liquid electrolytes that might leak during accidents.

Lead-acid batteries: They are the cheapest and the oldest sort of battery. Charging and functioning of them often result in the release of hydrogen, oxygen, and sulfur. They were used to power early versions of EVs in the 1970s.

Ultracapacitors: Useful in supplying extra power for acceleration and hill climbing, they may be utilized as supplementary energy storage in EVs since they can both quickly store and release energy — while protecting the primary batteries from overheating.

In recent years, there has been a lot of attention on alternative kinds of batteries. For example, Tesla last year chose the lithium-iron-phosphate (LFP) battery as their option for standard-range automobiles.

"From a chemistry aspect and a lifetime standpoint, they may last four to five times longer than a lithium-ion battery, and they're believed to be safer as well," said Josipa Petrunic, president and CEO of the Canadian Urban Transit Research and Innovation Consortium.

Tesla researchers have also been collaborating with Dalhousie University in Halifax to build a nickel-based battery that would be more robust than an LFP battery and potentially last 100 years or more in optimum circumstances.

The life span of an EV battery.

This depends on which batteries are being used and how you take care of them. Based on the existing EV industry, battery packs should come with an eight-year guarantee. However, Steve LeVine, editor of the Electric, a newsletter that specializes in electric cars and lithium-ion batteries, believes they last considerably longer.

When an EV battery pack declines to roughly 75 percent of its initial capacity, it is considered to be at the end-of-life stage.

"Car makers are anticipating the battery will decline, and it begins to fail on Day 1. But they want customers to have that 75 to 80 percent of that capability after around five to seven years," Petrunic added.

CHAPTER 4; The Future of EV Battery And its worldly impacts.

Electric vehicle batteries of the future might take numerous shapes. R&D departments throughout the world are experimenting with a variety of alternative technologies in a race to build the cheapest, lightest, most energy dense, and longest-lasting battery packs.

Here's what the EV batteries of the future may look like:

Battery Chemistry

In the future, the chemical makeup of electric vehicle batteries will undoubtedly alter.

Most electric vehicle batteries are constructed of a mix of metals including lithium, cobalt, and nickel. Launched in 1991, the lithium-ion battery is the most prevalent electric vehicle battery type, following on from its success in an assortment of consumer electronics such as smartphones and laptops.

When compared to the lead-acid 12V battery present in all traditional automobiles, lithium-ion batteries are more efficient and three times longer-lasting.

But they're not without their flaws. The most extensively used lithium-ion battery chemistry is lithium nickel manganese cobalt oxide (NMC) (NMC). But researchers have shown that lithium iron phosphate batteries (LFP), maybe both cheaper and safer, with higher thermal and chemical stability.

Researchers are also looking at lithium-sulfur batteries as a method to exclude pricey metals like cobalt and nickel from the mix.

Once manufacturing is scaled up, they might make EVs cheaper than petrol/diesel vehicles, and provide several additional advantages, including better energy density, and robust performance in temperatures as low as -30 degrees or as high as 60 degrees.

Lithium remains relatively uncommon, and this has pushed geologists to search and find new sources. But there are other viable alternatives to lithium. For example, sodium-ion batteries are fast-gaining popularity owing to their cheaper cost. One big negative, however, is that sodium is heavier and less appropriate for storing energy than lithium.

Solid-state batteries provide one of the most potential future alternatives to existing EV battery technology. Battery cells would employ a ceramic electrolyte instead of the organic liquids used in today's Li-ion batteries. This has huge consequences for how an EV battery functions. It considerably decreases the danger of fire and enables more energy-dense battery packs with a longer life expectancy and even quicker charging.

According to Toyota, we may anticipate solid-state batteries to be in production as early as 2025.

Battery range

If you've ever had an EV, or if you know someone who has, chances are you've heard of 'range anxiety. That's the sensation an EV owner gets when they don't know if they'll make it to the next charging site.

When the first mass-market EVs emerged over a decade ago, the median electric range was barely 68 miles. This statistic was (of course) only feasible on paper, under the correct driving circumstances, and (possibly) with the heater switched off.

To put that into perspective, on a chilly winter's day, a normal Scottish commuter could just about drive an 'average' EV from Glasgow to Edinburgh and back, providing they plugged in while at work and covered up for the trip.

Fast forward to 2020 and the median electric vehicle range was 259 miles - nearly 3.8 times as far!

So how far will EVs of the future go? Already, premium grade EVs like the Tesla Model S Plaid are within approach of the 400-mile threshold.

While firms like Tesla are looking to LFP batteries for a cheaper option, they are less energy dense. The same is true with sodium-ion batteries, which are predicted to be as much as 20% cheaper than LFP and have superior performance at lower temperatures. It's not unrealistic to predict an increasing disparity in the electric range between cheap and premium EVs in the future for this reason.

Several businesses are turning to solid-state batteries intending to unlock additional range. These batteries employ solid electrolytes (usual sodium) instead of the liquid electrolytes present in contemporary EV batteries. It may be many years before this technology hits the market, but when it does, experts say it might quadruple the range of existing EVs.

Structural batteries

There's a lot of discussion about 'energy density, but structural batteries might be another method to successfully cut weight while boosting the range of an electric car.

The essential premise is that batteries are not only effective for generating electricity, but also as a structural component. This might decrease the requirement for extra structural components, depending on the strength of the battery instead.

Currently, Tesla creates battery packs by assembling an array of cells into modules, which when placed together form a battery pack, which is then integrated onto the vehicle chassis.

At 'battery day' last year, however, Tesla stated that the battery pack for the new Model Y and Model S Plaid will constitute part of the structural framework of the car, employing a revolutionary honeycomb design for greater robustness. It would result in 370 fewer components, a 10% decrease in mass, and the prospect of up to a 14% greater range.

It's a brave approach in an industry that has been more inclined to cover EV battery packs, but it might permit the next great jump in EV range.

Lifespan

All EV batteries are rated for several 'cycles'. A battery cycle is a time it takes for a battery to be completely charged and discharged.

Over time, frequent cycles of charging and draining will deteriorate the battery. This lowers the maximum range of the automobile and the time between charges. Most manufacturers give a five to eight-year guarantee on their battery, yet it's predicted that the typical electric vehicle battery will last from 10 To 20 years before it has to be replaced.

How long an electric vehicle battery lasts may rely on several things, from the sort of environment the car is operated in, to the amount it is quickly charged, and how frequently it endures a 'deep discharge'.

In any event, the average lifetime of an EV battery might be poised to grow three-fold with the development of solid-state batteries.

Toyota intends to preserve more than 90% State of Health (SOH) of its solid-state batteries after 30 years of usage. In essence, this would imply an EV battery would only lose 10% of its maximum range over three decades, going well beyond the 'million-mile' battery revealed last year by Tesla.

Wireless charging

Wireless charging may seem like a bit of a gimmick on smartphones, but it might be ready to change EV charging.

In the UK, EV charging expert Charge is already trialing charging pad technology in anticipation that it would enable easier access to chargers for individuals lacking off-street access.

Wireless charging might prevent electric vehicle charging wire theft, decrease possible trip or accident dangers for other road or pavement users, and reduce unwanted clutter on the wayside.

The research involves Renault Zoes that have been modified with after-market induction kits, however, it's possible that if the trial proves successful, additional EV manufacturers may explore adding wireless charging technology in new EVs.

Sure, wirelessly charging while the vehicle is stationary is wonderful, but what about when driving?

The Smart road Gotland project is a 1.6km long 'electric road' connecting the airport and the town of Visby on the picturesque island of Gotland. Developed with Israeli business Electron, it might offer a smart solution for long-haul truck fleets by removing the need for massive, costly batteries and time-consuming charging breaks.
Should I purchase an electric vehicle now or wait for stronger batteries?

It's the question on everyone's lips: should I purchase an electric vehicle now, or wait? If there are so many fantastic breakthroughs on the horizon, perhaps it's better to simply sit patiently and wait for the next breakthrough in battery technology.

When it comes to expense, we can understand why you would be inclined to wait. But although EVs may now have a higher RRP than petrol or diesel vehicles, many are still eligible for a variety of UK Government awards for electric cars including the Plug-in Car Grant (PiCG) (PCG).

Surely future electric automobiles will be better for the environment too, you say? We've no doubt. But just because present EV batteries are anticipated to have a significantly shorter lifetime than EV batteries of the future, that doesn't imply they can't be recycled or reused.

As the initial generation of EV batteries slips into retirement, many are finding a new lease of life as batteries for electric forklifts, backup power supplies, and storage for solar generating installations. Meanwhile, automotive battery recycling means that manufacturers can reclaim valuable metals like cobalt from batteries that can no longer be utilized for alternative purposes.

While you may be able to achieve a longer range off-the-bat with future EVs, the UK's fast-increasing charging network means you'll never be too far from an electric vehicle ChargePoint. You can see for yourself with our electric vehicle charging station map!

And although future EV batteries may come with wireless charging built-in, there's no reason why inductive charging pads can't be retro-fitted if required, as we've seen with Char. guy's wireless charging experiment.

So go on… what are you waiting for?

CHAPTER 5; Conclusion

There's a lot to be thrilled about when it comes to EV batteries. New technology might mean it's soon feasible to travel between most areas of the UK on just a single charge, and that our EV batteries could soon be charged in a matter of minutes.

But it doesn't imply you should wait around till the next major idea. Investing in an EV today is a wonderful strategy to save expenses - and your carbon impact.

If you're shopping for your next vehicle, why not have a look at our lists of the cheapest electric cars and the best tiny electric cars for some inspiration?

And when you're ready to take the leap, compare electric vehicle leasing options with

Lease Fetcher to snag the lowest available pricing.

www.ingramcontent.com/pod-product-compliance
Lightning Source LLC
Chambersburg PA
CBHW050322220526
45465CB00005B/2096